Science Museum

Descriptive Catalogue
of the
Collection Illustrating

Fire Fighting Appliances

by K. R. Gilbert,
M.A., D.I.C.

Her Majesty's Stationery
Office London 1969

Contents

Illustrations

Preface

This catalogue of the Fire Fighting Appliances Collection lists the items held on inventory in January, 1969. The date of acquisition of an object appears as the first component of the inventory number, and these are listed on pages 42–45 together with the photograph negative numbers which should be quoted when ordering photographs.

Twenty-five of the exhibits are reproduced in colour in the Science Museum illustrated booklet on *Fire Engines*, which is obtainable from the Museum bookstall and from Her Majesty's Stationery Office at price 5s. 0d.

Where dimensions of drawings, etc. are given, the first dimension is the height and the second is the length. Where the original dimensions of an appliance were in British units, these are given followed by the approximate metric equivalents.

Most of the collection is on exhibition and of the small number of items in reserve some can be produced on request and others can be seen by special arrangement.

The numbers given in brackets in the following introduction refer to the catalogue entries.

Introduction

Fire is a phenomenon in which combustible materials, especially organic substances containing carbon, react chemically with the oxygen of the air to produce heat. Flame arises from the combustion of the volatile liquids and gases evolved and spreads the fire. Fire extinction is generally brought about by depriving the burning substances of oxygen and by cooling them to a temperature below which the reaction is not sustained. Applications of the principle of smothering are found in the use of carbon dioxide, carbon tetrachloride and certain other volatile liquids, which are employed in portable extinguishers and in fixed installations for dealing with special types of fire. Some of these substances are effective in sufficient concentration in inhibiting combustion although oxygen is still present. The use of foam, mechanically or chemically produced and consisting of bubbles containing air, will extinguish a fire by blanketing off the atmosphere. Foam is employed in attacking oil and petrol fires, where water would be ineffective because it would sink to the bottom of the burning liquid, whereas foam floats on the surface.

By far the most important extinguishant, however, by reason of its availability and general effectiveness, is water. It is more effective than any other common substance in absorbing heat, owing to its high specific heat and latent heat of vaporisation, thereby reducing the temperature of the burning mass. The steam produced also has a smothering action by lowering the oxygen content of the atmosphere near the fire.

A fire engine is therefore basically a transportable water pump and is in fact referred to by fire service personnel as a *motor pump* or, if it carries an escape, as a *pump escape*. Most fire brigades have at least one *turntable appliance*, which is also fitted with a pump, and several *trailer pumps*. The equipment of a brigade may also include a fire boat, hoselaying lorry, foam tender, emergency tender, control unit, breakdown lorry, canteen van, and other vehicles. All these are generically termed *appliances*.

The Romans had a large fire service, the Corps of Vigiles, equipped with pumps, buckets, ladders, fire hooks for pulling off burning roofs, and other tools. Double-cylinder force pumps are ascribed to Ctesibius of Alexandria, who is believed to have lived in the third century B.C., both by Hero and by Vitruvius. Hero stated that the pump was suitable for fire fighting and described the delivery pipe as jointed so that it could be directed up and down and from side to side. Vitruvius's description has been held to include the air vessel, which would have anticipated its introduction in the 17th century A.D., but this is now thought to be a misinterpretation. (See A. G. Drachmann. *The Mechanical Technology of Greek and Roman Antiquity*. 1963. pp. 155–7).

Three well-constructed bronze pumps of Roman date have been found, though they were not necessarily intended for fire fighting. They do not have air vessels. One from Bolsena is exhibited in the British Museum. With the collapse of the Roman Empire the fire pump fell into disuse and was only re-invented about the beginning of the 17th century. The earliest known illustration of a fire engine is in a Book of Machines by Heinrich Zeising, published in 1612. It shows a two-cylinder force pump, with swivelling delivery branch, mounted in a tank on wheels. There was no air vessel. Another engine was mounted on a sled, as was a similar engine depicted by de Caus in 1615. The syringe or squirt (1) had already been in use for some time and one is illustrated in 1556 in Agricola's *De Re Metallica,* hanging on a wall with buckets, fire hooks, and a sledge hammer.

The introduction of the air vessel to maintain the delivery pressure was due to Hans Hautsch of Nuremburg in 1655. The water passed from the pump cylinders through valves into a vessel containing air from the bottom of which led a tube to the delivery nozzle. As the water level rose above the outlet, the air was compressed and maintained the flow between strokes, so that the jet of water was continuous instead of intermittent. This stage in the evolution of the fire pump is represented by a contemporary model (5). Van der Heiden in Holland brought about the next important improvement in fire fighting by his introduction in 1673 of the fire hose, which enabled a close attack on the fire to be made instead of the long shot from the engine (94). At first the cistern of the engine was connected by flexible hose with a canvas trough in a wooden stand, which was kept filled with water brought in buckets, but van der Heiden later introduced the wired suction hose, which was dipped into a canal, nowhere far away in Amsterdam. The Dutch engines were introduced into England in 1688 on the accession of William and Mary.

In the 18th century in England there were several manufacturers of fire engines, of whom the best known was Richard Newsham (6), who took out patents in 1721 and 1725. He placed the levers at the sides instead of at the ends to enable more men to work at it and added treadles to reinforce the pumping effort. The rocking motion imparted an up and down motion to the pistons by means of cast-iron segments and link chains. He also provided a three-way cock so that the engine could either work from its cistern or by suction and the cistern could be drained after use. Towards the end of the 18th century larger engines were built. They were arranged for horse traction and mounted on springs and the treadles were omitted (14). Charles Simpkin, later a partner in the firm Hadley, Simpkin, and Lott, the predecessor of Merryweather & Sons, in 1792 patented the metal valve to be used instead of the leather valve. The valves were placed in separate valve chambers to make them more accessible, instead of being within the cylinders and air vessel (18). Another improvement was to use folding handles, so that many men could pump and nevertheless the machine remained conveniently compact for travelling. The manual engine continued in use during the 19th century (**17**).

Fire fighting in the 17th and 18th centuries was legally the duty of the parish authorities. By an Act of Parliament of 1707 the church-wardens were obliged under penalty to ensure that a large engine, a hand engine, and a leather pipe were kept in good repair; and the keepers of the engines were encouraged by rewards to be first, second or third to arrive at a fire. In London after the Great Fire of 1666 the Fire Office for fire insurance was set up and competition soon appeared with The Friendly Society in 1684 and other companies. In order to protect properties at risk they formed private fire brigades. The firemen wore splendid uniforms and there was keen rivalry to demonstrate the protection they could provide for policy holders (**12, 15, 98**). They were in fact a far more effective force than that provided by the parishes and the parish engines ceased to be of much account. In 1824 the insurance companies with the help of the city authorities established a unified fire brigade in Edinburgh under the command of James Braidwood and in 1832 the insurance brigades in London amalgamated to form the London Fire Engine Establishment with Braidwood as Superintendent. In 1866 the Establishment became a public responsibility as the Metropolitan Fire Brigade, which was renamed the London Fire Brigade in 1904.

Fire Brigades were the responsibility of local authorities until 1941, when the National Fire Service was formed; but in 1947 the Service was disbanded and returned to local authority control, not however to the 1,441 pre-war authorities, but to 146 counties and county boroughs, which are units big enough to maintain well-equipped brigades. The Secretary of State for Home Affairs retained responsibility for securing the efficiency of Fire Services, which is exercised through the Civil Defence and Fire Service Department. Subsequent amalgamations have reduced the number of brigades to 132. In addition there are eleven brigades in Scotland and two in Northern Ireland. Altogether there are about 47,000 fireman in the public service with 3,600 appliances at their disposal.

During the 19th century there were many volunteer brigades and particularly notable is the Royal Society for the Protection of Life from Fire, which was formed in 1836 and established fire-escape stations in the Metropolis (**44–48**). Private fire brigades were maintained to protect country houses, works, and institutions (**17, 21**); and more than 500 firms now have their own appliances with trained firemen in charge.

The application of steam power to the fire engine came late and the first steamer, which was horse drawn, was designed by John Braithwaite in 1829 (**22**). It was used with great success at the Argyle Rooms Fire, but was not generally adopted, because the output was thought to be too large for the street mains to supply and for other reasons. It was only in 1860 that London used its first steam fire engine, supplied by Shand, Mason & Co. This firm, which was a successor of Tilley (**14**), and

Merryweather & Sons were the leading manufacturers of fire-fighting equipment during the 19th century and later. Merryweather & Sons Ltd. eventually acquired Shand, Mason & Co. in 1922. In 1863 a competition was organised to determine the best design, which was won for the *Sutherland* (**24**), which is probably the oldest steam fire engine extant. In the same year Shand, Mason & Co. brought out their vertical engine (**25**), which became the most generally used type especially for smaller appliances. The double-cylinder vertical engine was introduced in 1889 (**32**).

The self-propelled steam appliance was invented by Hodge in 1840 (**23**) and this rather heavy type of appliance became common by the end of the century, but was superseded by the motor fire engine, the first of which was built in 1904 (**36**).

The motor fire engine continued for many years to be an open vehicle, following the practice of the horse-drawn manual in which the firemen stood holding on to a rail or sat on longitudinal benches (**37**), but in 1930 the enclosed body was introduced to give protection to the crew (**39**). Although the early pumps were reciprocating, the high speed of the internal combustion engine made it practicable to use the simpler and more compact centrifugal pump. At first petrol engines were used for motor fire engines, but many modern appliances in common with other heavy vehicles are diesel-powered.

The light-weight portable pump (**42**), which may be used in places inaccessible to a wheeled vehicle, may be driven by a petrol engine or by a gas turbine. Several designs of turntable ladder were patented from 1888 onwards and such appliances were in use early in the 20th century. Manchester had the first horse-drawn turntable appliance in Britain in 1904. Mounted on a vehicle the ladder can be extended, elevated, and swung in a complete circle on its turntable. Fully motorised by 1908, the 'T.L.' can be used both as an escape and as a water tower from which a jet can be directed into a tall building from a height of 100 feet (**40**).

Prompt action is most important when dealing with a fire and may indeed make recourse to the large appliances at the disposal of a brigade unnecessary. Portable fire extinguishers contain a quantity of extinguishing medium and a means of expelling it under pressure through a nozzle. The first extinguisher was invented in 1816 by Captain George Manby. It consisted of a four-gallon copper vessel holding three gallons of water, in which salts thought to improve its extinguishing properties were dissolved. The outlet pipe closed by a stop-cock reached inside to near the bottom of the vessel. The space above the liquid contained air compressed by a pump, when charging the extinguisher. Manby's extinguisher is illustrated in action on the front cover.

The gas pressure may also be produced by carbon dioxide generated chemically as patented by François Carlier in 1862. The addition of liquorice extract to produce a foam, which would exclude the air from a burning oil surface, was patented by H. H. Lake and A. G. Laurent in 1906. The use of powder as an extinguishant was patented by B. Vorwerk in 1903, but some fifty years elapsed before powders with satisfactory flow and extinguishing properties were developed. Finely divided powders exclude air by covering the surface of the burning substance and there may also be chemical action, a factor which makes the choice of powder an important consideration. Carbon tetrachloride which is used in motor-vehicle fires and on electrical fires, because it is a non-conductor, was patented as an extinguishant by E. M. Davidson in 1909.

Prompt and automatic fire protection of buildings can be provided by a sprinkler system. The forerunner of this was the proposal in 1797 of Sir Samuel Bentham, Inspector General of Naval Works, to protect buildings in Portsmouth Dockyard by placing a water tank on the roof connected by a system of pipes to hydrants. In 1806 John Carey patented a system of perforated pipes to which water was admitted through valves, which were held shut by weighted string and opened when the string burnt. Stewart Harrison in 1864 invented a practical system in which the sprinklers acted individually and released water on the fusing of a plug. His proposal was not adopted, but a similar system was patented and put into practice in New Haven, Conn. in 1874 by Henry S. Parmelee. Numerous sprinkler designs were then patented, the most successful of which was that of Frederick Grinnell, whose first sprinkler was invented in 1881 (**71**). The value of sprinkler protection is attested by the substantial reduction in fire insurance premiums which their installation attracts.

The notification of fires to Fire Brigade Headquarters by telegraphy was started in 1849 in Berlin and a street fire alarm system was installed in Boston, Mass. in 1852, which was subsequently developed by the Gamewell Company (**84**). These systems depended on a watchman or a passer-by giving the alarm, but modern fire alarm systems depend on the use of a device able to detect the presence of fire. Detectors sensitive to temperature rise, to smoke, or to ionisation operate electric relays, which set off audible alarms and notify the brigade or a fire watch centre. (**85–92**). A bimetallic fire detector responsive to temperature rise was patented in 1852 by D. L. Price, but automatic fire alarms only became common half a century later. The actuation of a sprinkler system (**70**) is also arranged to give the alarm, to call firemen who will cut off the water, when satisfied that the fire is out.

Hand pumps

1 Fire Squirt c. 1750

(See figure 1)

This fire squirt or syringe, which was made about 1750, was one of three originally kept ready for use in the Parish Church of Saint Dionis-Back-church, Fenchurch Street, London. The barrel, nozzle, and side handles are of cast brass. The handles were probably used for attaching the squirt to a sling on the back of the fireman for transport, and they were used by two men for directing the jet and holding the squirt while the forcer was pressed in by a third. The brass piston is packed with hemp and is fitted with a wooden piston rod. The bore of the barrel is $2\frac{1}{2}$ in (6·35 cm) and of the nozzle $\frac{1}{2}$ in (1·57 cm); the stroke is 18 in (46 cm).

2 Hand Pump

(See figure 1)

The addition of an air vessel and valves to the hand syringe, by which it was converted into an efficient hand pump, was due to William Baddeley prior to 1844. It was adopted by the London Fire Engine Establishment in 1848; and, as the *London Brigade* pump manufactured by Shand, Mason & Co., it continued in general use until it was superseded by the stirrup pump.

The single-acting pump discharges water through the delivery valve into the space between the cylinder and the outer casing. On the down-stroke some of the water passes through the discharge pipe, but some of it rises above the opening into this pipe, so compressing the air in the casing. The compressed air maintains the flow on the up-stroke.

The pump was supplied with a covered pail, which was kept filled with water ready for use.

A sectioned pump is also exhibited.

3 Stirrup Pump

(See figure 1)

This form of double-acting pump was made in large numbers for fighting the fires caused by incendiary bombs in the 1939–45 war. It was designed so that it could easily be used by one man. The barrel is placed in a bucket of water and the pump held steady by placing one foot on the stirrup, thus leaving the hands free to work the plunger and to hold the nozzle. The construction is very simple and the pump has only three moving parts: the plunger rod, a ball in the foot valve, and the ball which forms a non-return valve in the piston. A slide mechanism in the nozzle provides the alternatives of a $\frac{1}{8}$ in (32 mm) jet throwing 30 ft

(10 m) or a fine spray carrying 15 ft (5 m). A length of oiled string is used for packing at the top of the cylinder. On the up-stroke the piston valve closes and the foot valve opens. Water above the piston is forced through the discharge pipe and water enters below through the foot valve. On the down-stroke the foot valve closes and the piston valve opens. Water enters the space above the piston and the discharge continues as the water in the barrel is displaced by the plunger rod.

A sectioned pump is also exhibited.

4   Primitive Japanese Fire Pump

This form of hand-pump was in general use in Japan in 1860 for extinguishing fires. The lower end was placed in a bucket of water, and the inclination of the intermittent jet produced on working the handle was adjustable by a swivelling joint in the branch pipe.

The barrel is formed by boring a hole 3 cm diameter through a rectangular post 1 m long, in which another hole is bored parallel with it up to the junction for the delivery pipe. The lower end of the barrel is fitted with a foot valve in the form of a sheet metal clack, loosely pinned to a wooden seating ; and a similar fitting secured in the side passage, at a point where there is a hole connecting with the pump barrel, forms the delivery valve. The piiston is on a rod resembling a plunger, provided with a cross handle. The jet delivered is 5 mm diameter.

A sheet of thin metal is nailed over the bottom end of the pump. This is perforated with many small holes so as to act as an inlet strainer.

Manual fire appliances

5 Manual Fire Engine *c.* 1680
(See figure 2)

This model may have been made by a 17th century fire engine maker to show to customers, as the provision of full-size handles out of scale with the model seems to indicate ; and it is believed to be the earliest model of a fire engine extant.

It consists of two vertical pumps in a metal cistern mounted on a sled. There is no suction inlet and the cistern had to be filled with water brought in buckets. The pumps alternately force water into a large copper air vessel placed between them. When the water level rises above the outlet pipe going to the nozzle, the air is compressed and ejects the water in a continuous stream. The delivery pipe has two swivel joints in different planes, so that the jet may be pointed in any direction.

6 Newsham Fire Engine 1734
(See figure 3)

Richard Newsham (*d.* 1743) of London, in 1721 and 1725 patented 'a new water engine for quenching and extinguishing fires'. Newsham's engine, while it probably owed much to the earlier Dutch engines, was a great improvement on previous machines and was generally adopted. This example, which is dated 1734, has two single-acting pump barrels 4·5 in (11·5 cm) diameter, 8·5 in (21·5 cm) stroke, and a tall air vessel to secure a continuous discharge. The pumps are placed in a tank which forms the frame of the machine, and the water to be pumped was brought in buckets and emptied into the tank, but the suction inlet to the tanks is provided with a two-way cock, by which the pumps can be arranged to draw either from the tank or through a length of suction hose.

Leather hose for 'conveying water to and from fire and other engines' was patented in 1676, and Newsham used it, connecting his suction hose at the base and the delivery at the top of the casing enclosing the air vessel. The engine could also be used with a jet pipe or branch connected with the outlet of the pump by means of a swivel joint or goose neck and this one, which came from Dartmouth, is so exhibited, fitted with the branch of the Phillips engine (No. 8) and a goose neck copied from an engine in the possession of the London Fire Brigade.

The pumps were worked by men at the long cross handles, but in addition two treadle boards were provided near the centre of the machine, upon which several more men stood and assisted the pumping by throwing their weight on the descending treadle.

On the front panel of the engine, but protected by a sheet of horn and a door, are directions for keeping the engine in order. Below this are the coats of arms of the Borough of Dartmouth and of Walter Cary, Member of Parliament for the Borough and presumably the donor of the engine, and the date.

In 1966 the engine was repainted in its original colours, except for the coats of arms and date which were cleaned and retouched.

7 The Cornhill Fire
(*See figure 4*)

This print, 18 cm x 41 cm, shows Newsham's engine at work at the fire of 1748 in Cornhill, London. The scene is flanked by two firemen wearing the badges of the Sun and Phoenix Insurance Companies.

8 Manual Fire Engine 1766
(*See figure 5*)

This is a small compact machine which differs in several respects from Newsham's. It was made by Samuel Phillips and was presented by 'Benn. Way, Esqr., to the Parish of Denham, Bucks, 1766,' as an inscription testifies.

The engine has two single-acting pump barrels, 4 in (10 cm) diameter by 8 in (20 cm) stroke, with a tall air vessel between them. These are placed in their casing longitudinally on the centre line of the reservoir instead of transversely as in Newsham's engine. The piston-rods are guided and driven by forked connecting rods from the rocking shaft. The pump handles are also placed transversely, thus increasing the stability of the machine when operated, and are arranged to fold up so as to reduce the lateral space required for storage.

9 Trade Card of John Bristow
(*See figure 6*)

The trade card of John Bristow, Ratcliff Highway, London, shows one of this manufacturer's manual fire engines. On the back is a bill to the Parish of St. Laurance Jury dated Dec. 25th, 1773 :

'To Men's Attendance Once Every Quarter in taking to pieces Cleaning Oyling & Keeping in Order the Engine & Leather pipes from Ladyday 1773 to Xmas 1773 at 4 p. anno £3.0.0'.

10 Hadley's Broadsheet

Nathaniel Hadley was in business as a fire-engine maker in Long Acre, London, from 1769 to 1790 and was a predecessor of Merryweather & Sons. This broadsheet shows a fire-fighting scene.

11 Fire at the Albion Mills
(*See figure 7*)

This coloured engraving, 24 cm x 28 cm shows fire engines at work at the fire which took place on the 3rd of March 1791, at the Albion Mills, Blackfriars Bridge, London. It is one of the engravings from Ackermann's *The Microcosm of London*, 1808.

B

12 Sun Fire Office Fireman
(*See figure 8*)

This coloured aquatint, 31 cm x 23 cm, from W. H. Pyne's *The Costume of Great Britain*, 1808, is dated 1805 and shows a fireman of the Sun Fire Office. A combined treadle and manual engine is depicted in the background.

13 Manual Fire Engine
(*See figure 9*)

This hand-drawn machine, made about 1820, or a little earlier, by Simpson & Sons, was formerly used at Windsor Castle. It is similar in general design to the Newsham engine, but is larger and embodies improvements in detail.

The body is a long narrow box, 10ft (3 m) long by 2·25 ft (0·7 m)wide, mounted on four cast-iron wheels. The two gunmetal pumps have barrels 8·5 in (21·5 cm) diameter and are placed in a casing in the rear of the machine. The valves are placed in separate valve boxes apart from the pump barrels and air vessels, and are thus rendered easily accessible, as patented by Charles Simpkin in 1792. A large copper air vessel is placed in front of the pump casing and is connected to the pressure side of the pump. The water can be drawn through a two-way cock either from the reservoir formed in the body of the machine itself or through a suction pipe at the rear. The outlet is at the front of the machine. The side handles, which are long enough to enable as many as sixteen men to work the machine, oscillate the main shaft, which is supported on three heavy gunmetal bearings, fixed on cast-iron brackets. The guided piston rods of the pumps are driven by connecting rods attached to short arms on this shaft, giving a stroke of about 7 in (18 cm).

14 Manual Fire Engine, early 19th Century
(*See figure 10*)

This contemporary model to about ¼ scale shows the kind of horse-drawn manual fire engine used in the early part of the nineteenth century. Although horse-drawn manual fire engines had been in use earlier, their development was slow until the beginning of the nineteenth century.

The machine represented, which was made by W. J. Tilley, 1820–51, one of a line of fire engine manufacturers and the immediate predecessor of Shand, Mason & Co., was clearly a practical machine, and, if compared with the model of the later standardised and very successful London Brigade manuals of the latter part of the century (No. 18), it will be seen that its construction is generally very similar and the improvements in the later machines are mainly in small details and additional refinements. The pump has two vertical single-acting cylinders with the valves in separate valve boxes for greater accessibility as patented by Charles Simpkin in 1792. The suction hose is fitted to the back of the machine, while provision is also made for supplying water to the machine by emptying buckets into a trough at the back. Two screwed couplings are provided for the delivery hose, one on each side of the machine.

15 London Fire Engines
(See figure 11)

Horse-drawn fire-engines of the County, Westminster, and Phoenix Insurance Companies on their way to a fire are depicted in this coloured aquatint, 51 cm x 76 cm, after the original painted by James Pollard in about 1825.

It is entitled *London Fire Engines. The Noble Protectors of Lives and Property*.

The parish engine is to be seen in the background and another engine is already at work at the fire.

16 Manual Fire Engine 1863

This model (Scale 1:8) represents the *Paxton* horse-drawn Brigade manual fire engine made by Merryweather and Sons, who were awarded a prize for it at the 1851 Exhibition.

The engine was manufactured in two sizes and its lightness made it specially suitable for country use.

17 Manual Fire Engine 1866
(See figure 12)

This is the smallest of the *Brigade* horse-drawn fire engines introduced by Merryweather & Sons in 1851 and known as the *Paxton*. It has pumps 6 in (15·2 cm) by 8 in (20·3 cm) stroke, and with 22 men to work it, would deliver 100 gallons (450 l) of water per minute to a height of 120 feet (40 m).

This engine was purchased in 1866 by the Duke of Portland for his Welbeck Abbey Estate.

Following an accident sometime subsequent to 1880 the engine was repaired and the opportunity taken to add brakes and spring clips to the sway bars, both of Shand, Mason & Co, pattern.

It is exhibited with the figure of a fireman wearing, except for the helmet, which is shown separately (No. 99), the uniform of the Duke's private fire brigade.

The appliance was repainted in 1966.

18 Manual Fire Engine *c.* 1881
(See figure 13)

This sectional model to scale 1:4·8 shows the construction of a large horse-drawn manual fire engine made by Merryweather & Sons for the London Metropolitan Fire Brigade. In 1851 an improved and more or less standardised machine known as the *London Brigade* type was introduced and in the following years, with improvements added from time to time, was much used. The model shows the machine as it was in about 1881.

The body is of timber and carried on large road wheels with a forelocking carriage. It is mounted on springs throughout, and provided with a powerful hand brake, and generally arranged for rapid hauling by two horses. The roof of the body is hinged so as to give access to the interior, in which the hose and other accessories are conveyed, while the roof also served as a seat for the firemen and is furnished with two cross hand-rails for enabling them to retain a firm hold when in motion. The tops of the long side boxes, which contain the two lengths of suction hose, also served as footboards.

The pumping machinery consists of two vertical single-acting pumps driven by links from the horizontal shaft connected with the handles by levers outside the framing. The barrels are of cast brass, with pistons made of two circular pieces of brass, each put into a leather cup and bolted together. The cups are oiled to make a good seal with the cylinder walls. The valves are hinged brass plates, truly ground to fit the circular brass orifices on which they fall and are readily accessible by two covers which are held down by set screws. The suction can be taken in through suction hose or, when working close to the main, the hydrant delivers directly into the chamber containing the pumps, this arrangement being the same as in Newsham's fire engine (No. 6). The delivery passes out through a branched pipe having a union on each side of the engine, and on this pipe is a large copper air vessel to secure a uniform discharge.

The model is exhibited with a full set of tools and accessories.

19 Manual Fire Engine c. 1881
This model to scale 1 : 4·8 represents the same engine as No. 18, but is not sectioned.

20 Wheeled Fire Pump
(See figure 14)

This type of appliance was manufactured in the late 19th and early 20th century for use in large buildings and was known as a corridor engine. It consists of a pump operated by a hand lever and installed in a cistern, which stands on two wheels and two feet and has a pair of handles for wheeling it about. The container was kept filled with water and the appliance could be used by two persons, one to work the pump and the other to direct the jet; or, if necessary, by one person alone.

21 Manual Fire Engine 1898
(See figure 15)

This is the small *Factory* engine built by Merryweather & Sons. Owing to its simplicity and consequent easy maintenance and low cost, the small manual engine was retained for use in country houses, institutions, and small factories, long after the public brigades and larger establish-ments had adopted the steam and later the motor fire engine. This engine was supplied to the Office of Works for use in Dorchester Prison. It was hand drawn and was worked by twelve men. It was repainted in 1966.

Steam fire appliances

22 Braithwaite and Ericsson's Fire Engine 1829–30
(*See figure 16*)

This coloured drawing represents the first steam fire engine, which was constructed in London by Messrs. John Braithwaite and John Ericsson in 1829–30.

Although the engine was worked gratuitously at several fires in London with great success, it met with determined opposition. Three or four similar engines were, however, built for Liverpool and abroad, but it was not until 1852 that steam was adopted in London, when a floating engine was converted to steam working.

The engine had a boiler and two direct-acting steam pumps of $6\frac{1}{2}$ in (16·5 cm) diameter and 16 in (40·6 cm) stroke. It was mounted on wheels for horse traction. The fire box was water-jacketed and was provided with a forced draught by a mechanical bellows and the waste gases issued from the funnel behind the driver's seat. The engine threw 150 gallons (680 l) of water per minute to a height of 90 ft (30 m). The drawing measuring 37 m x 65 cm, is to 1 : 16 scale, and is signed G.Pooley and dated 1869. The mount is signed by the inventor John Braithwaite.

23 Steam Fire Engine 1840
(*See figure 17*)

The coloured drawing shows the first self-propelled fire engine, which was also the first steam fire engine to be built in America. It was constructed by P. R. Hodge (No. 101) at New York in 1840.

When the scene of operations was reached, one end of the engine was raised till the driving wheels were off the ground, when they acted as fly-wheels. The horizontal pumps were placed tandem with the steam cylinders, and were directly driven by the piston rods produced backwards. The artist, date, size, and scale are the same as for No. 22, but the mount is signed by Paul R. Hodge.

24 Steam Fire Engine 1863
(*See figure 18*)

This engine, the *Sutherland*, was built by Merryweather & Sons and won the first prize for large steam engines at the international competition at the Crystal Palace in 1863. It is believed to be the oldest steam fire engine still in existence.

Although a steam fire engine had been made by Braithwaite and Ericsson in 1829, this type of machine was not developed in England until after the middle of the 19th century, In 1858 Shand, Mason & Co made their first steam fire engine which was sold to Russia. In 1860 the London Fire Engine Establishment used its first land steam fire engine. Soon after this time interest in such machines developed and competitions were organised in order to discover the best designs.

The machine shown belongs to this early period of rapid practical development. It has a pair of horizontal pumps working direct off the piston rods of the two steam cylinders, so that there is no crank shaft or flywheel. Under test it proved capable of maintaining a steady jet 160 to 170 ft (50 m) high through a 1·525 in (38·8 mm) diameter nozzle. It was purchased by the Admiralty and used for 27 years at Devonport dockyard.

In 1905 it was taken out of service, but in 1918 it was once again in use, and finally in 1924 was placed in this Museum. Since its construction it has undergone a few alterations. The chief change has been the substitution of the present boiler for the original Field tube boiler. Some of the woodwork has also been renewed and it has been repainted in its original colours.

25 Steam Fire Engine c. 1866
(See figure 19)

This model (Scale 1 : 8) represents an early version of the vertical style of engine introduced in 1863 as the *London Brigade Vertical*. This appliance was built by Shand, Mason and Co and is the first example in a fire engine of the short-stroke high-speed type of engine.

The vertical double-acting steam cylinder is placed directly over and concentric with the pump cylinder fitted with bucket and plunger. The plunger is connected directly to the steam piston by two rods and to a crank by a connecting rod joined to the bottom of the plunger. The crankshaft carries a flywheel and operates the valve gear.

26 Steam Fire Engine c. 1876
(See figure 20)

This model (Scale 1 : 4·8) represents a later example of the *London Brigade Vertical* type of engine which was introduced by Shand, Mason & Co in 1863.

This machine is carried by large wheels on mail-coach axles, with a forelocking carriage, and is generally arranged for rapid travelling, two horses being employed. The delivery hose is stored in a central box, which also contains a tank for the boiler feed water. The two lengths of heavy suction hose are carried one on each side beneath the firemen's footboard, and a coal bunker under the fore carriage holds a supply of fuel conveniently near the furnace door. The box seat is utilised as tool store, and two short branch pipes with nozzles are carried, one each side of the box, and two long ones over the hose box. The pole terminates in a special fitting, by which a horse that has fallen can be quickly released, and the back wheels have a powerful lever brake.

The pumping machinery consists of a vertical cylinder 7·5 in (19 cm) diameter, 7 in (17·8 cm) stroke, and a double-acting bucket and plunger pump with an 8 in (20·3 cm) bucket and a 6 in (15·2 cm) plunger. The piston is directly connected to the plunger by two piston rods, which permit the horizontal crankshaft to be placed on the centre line of the engine, a return connecting rod from the interior of the plunger connecting the reciprocating parts with the crank pin. A full stroke is thus secured and a flywheel introduced which carries the engine over its dead centres. An overhanging eccentric works a slide valve, and also a feed pump attached below in line with the valve-rod. The suction chamber of the pump is directly below the barrel, and the delivery takes place above, two sluice valves being arranged to direct the water into two lengths of delivery hose. A spring-loaded relief valve permits the return of water from the delivery into the suction valve-chamber if the pressure exceeds the limit for the hose. A large copper air-vessel is placed on both the suction and delivery boxes, to prevent any concussion in the pipes, and a pressure gauge in front of the boiler registers the pressure in the hose.

Steam at 100 lbs pressure per sq in (7 kg per sq cm) is supplied by a vertical boiler, internally fired and having the fire-box crossed by rows of inclined water tubes. The boiler shell is in two lengths, bolted together, to render the fire-box accessible for cleaning and repairs. The boiler is provided with two spring-loaded lever safety valves, an injector, steam whistle, and a jet in the funnel for drawing up a poor fire.

The engine indicates 30 hp delivers 350 gallons (1,600 l) per minute and weighs rather under 1½ tons (1,500 kg). Steam is raised to full pressure in six minutes from the time of lighting the fire. A full set of tools and accessories is shown with the model.

27   Steam Fire Engine *c.* 1876

This model, except for minor details, is the same as No. 26.

28   Steam Fire Engine *c.* 1876

A sectional drawing, 67 cm x 118 cm, of No. 27, by Thomas Coates, the maker of the model.

29   Steam Fire Engine 1885

(*See figure 21*)

This model (Scale 1 : 4·8) shows the construction of a large Merryweather steam fire engine such as was used on docks and other large areas of warehouse property. This engine is a development of the *Sutherland* and embodies improvements patented by Messrs H. Merryweather and C. J. W. Jakeman in 1880.

The pumping machinery consists of two horizontal steam cylinders 7 in (17·8 cm) diameter by 7 in stroke, and two double-acting horizontal piston pumps 5 in (12·7 cm) diameter The steam pistons are directly connected with the water pistons by common piston rods. Nearly all the engines of this type, though fitted with interconnected cranks, were without flywheels, and the arrangement shown in the model, which has a flywheel, was only used for one engine. A large copper air vessel is placed on top of the delivery chamber, and a small one on the suction chamber to give a steady flow in the pipes.

The engine delivered 450 gallons (2,000 l) per min, and with a 1·5 in (38 mm) nozzle the jet reached a height of 190 ft (60 m). Steam was raised to the full working pressure in ten minutes from the time of lighting the fire.

30   Steam Fire Engine 1885

A sectional drawing, 70 cm x 121 cm of No. 29, by Thomas Coates, the maker of the model.

31   Steam Pumping Unit 1894

This is a model of the single-cylinder pumping unit fitted by Shand, Mason & Co, in their 260 gallon (1,180 l) horse-drawn steam fire engine.

32   Steam Fire Engine 1894
(See figure 23)

This is an example of the *Double Vertical* engine introduced by Shand, Mason & Co, in 1889, which remained the standard type of horse-drawn fire engine until it was superseded by the petrol-motor fire engine.

This engine, which came from Southgate, Middlesex, is one of about twenty made to the original pattern and is of the most popular 350 gallon (1,600 l) per minute size. It has two double-acting steam cylinders working directly on to two double-acting pumps placed vertically below them. At each end of the crank-shaft are eccentrics for working the slide-valves of the steam cylinders. The cranks are set at right angles to allow starting in any position, and small fly-wheels to ensure smooth running are fitted at each end of the crank-shaft.

The furnace door is placed on the side towards the front instead of at the back of the boiler as in earlier engines. This arrangement permitted the stoker and the engineer to work without mutual interference and by allowing the machinery to be placed lower, increased the stability of the engine when travelling.

There is an inclined water-tube type of boiler, and steam could be raised from cold water to working pressure in less than ten minutes while the engine was travelling to a fire.

The appliance was repainted and restored in 1966.

33   Steam Fire Engine 1902
(See figure 23)

This was the last horse-drawn fire engine in the London Fire Brigade. The last occasion on which a steam fire engine was used in London was at Peckham in 1917.

This appliance was built by Merryweather & Sons, Ltd, and was specially designed for use in hilly districts. It is similar to the large engines introduced in 1893, but is much lighter, weighing 27 cwt (1,370 kg).

1 Fire squirt (**1**),
 London Brigade
 pump (**2**), stirrup
 pump (**3**), and
 leather fire bucket
 (**93**)

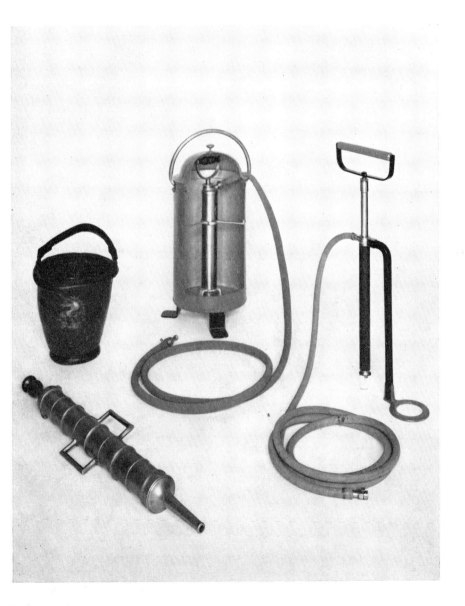

2 Manual fire engine,
   *c.*1680 (Model) (**5**)

3 (*right above*)
   Newsham manual
   fire engine, 1734 (**6**)

4 (*right*) The Cornhill
   Fire, 1748 (**7**)

A Perspective View of part of the RUINS of the late dreadful FIRE which happened in Cornhill, on March 25, 1748.

7   Fire at the Albion
Mills, 1791 (**11**)

5   (*left above*) Phillips
manual fire engine,
1766 (**8**)

6   (*left*) Bristow's
trade card, 1773 (**9**)

8 (*left*) Sun Fire
Office fireman, 1805
(**12**)

9 (*left below*) Simpson
manual fire engine,
c.1820 (**13**)

10 Tilley manual fire
engine, early 19th
century (Model)
(**14**)

12 (*right*) Merryweather
    manual fire engine,
    1866 (**17**)

11 London fire engines,
    *c.*1825 (**15**)

13 Merryweather
manual fire engine
(Sectional model)
(**18**)

14 Corridor engine (**20**)

15 Merryweather
Factory manual fire
engine, 1898 (**21**)

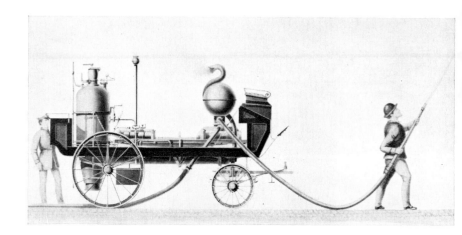

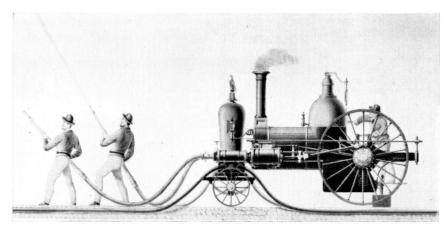

16 (*top*) Braithwaite
    and Ericsson's
    steam fire engine,
    1829–30 (**22**)

17 Hodge's steam fire
    engine, 1840 (**23**)

18 Merryweather steam
fire engine, the
Sutherland, 1863 (**24**)

19 Shand Mason
*London Brigade
Vertical* steam fire
engine, *c.*1866
(Model) (**25**)

20 (*below*) Shand
Mason *London
Brigade Vertical*
steam fire engine,
*c.*1876 (Model) (**26**)

21 (*above*) Merry-
weather steam fire
engine, 1885
(Model) (**29**)

22 Shand Mason
Double Vertical
steam fire engine,
1894 (**32**)

23 Merryweather steam
fire engine, 1902
(**33**)

24 *(top)* Steam fire
float, the *Fire Queen*,
1884 (Model) (**35**)

25 Merryweather motor
fire engine, 1904
(**36**)

26 Leyland motor fire
engine, 1936
(Model) (**37**)

27 Dennis dual purpose
appliance, 1936 **(38)**

28 Leyland limousine
fire pump, 1936
(Model) (**39**)

29 Leyland-Metz
turntable appliance,
1936 (**40**)

30 Leyland-Metz
turntable appliance,
1936 (**40**), fully
elevated and ex-
tended at the rear of
the Science Museum

31 Simon Snorkel
hydraulic platform,
1965 (Model) (**41**)

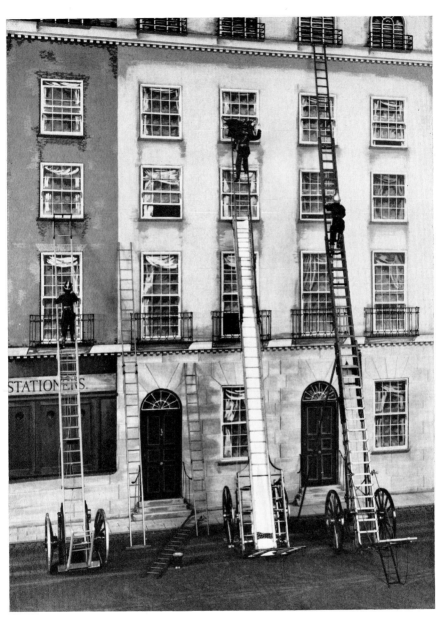

32 (*left above*) Coventry
    Climax portable fire
    pump, 1966 (**42**)

33 (*left*) Fire boat, the
    James Braidwood,
    1939 (Model) (**43**)

34 Fire escape models,
    From left to right:
    extending ladder,
    1836 (**46**); hinged
    ladder, 1836 (**47**);
    fly ladder, 1836 (**45**);
    extending ladder,
    1908 (**49**)

35 Fire scene with fly-
ladder escapes in
use, c.1836 (**48**)

36 (*right above*)
Wivell's fire escapes,
1827–35 (**48**)

37 (*right*) Mr Wivell on
a fly-ladder escape,
1836 (**48**)

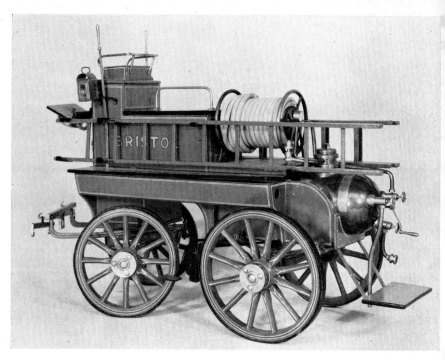

38 Chemical fire engine,
1902 (Model) (**55**)

39 *(right)* Firemark of the
Sun Fire Office with
photograph of the
entry in the Policy
Register, 1720 (**98**)

40 (*far right*) Sprinkler
installation : 2 in.
wet valve
(Sectioned) (**70**)

41 Illustration from Jan
van der Heiden's
book showing the
old and the new
engines (**94**)

The engine is vertical, double-acting, and has two cylinders, 5 in (12·7 cm) diameter by 3·5 in (8·9 cm) stroke. The boiler is of the vertical water-tube type, originally oil-fired, but converted to burn coal.

34 Fire Scene in London *c.* 1900

A water-colour painting, 45 cm x 61 cm, by H. B. Thomas depicting steam fire engines in action at a fire.

35 Fire Float 1884

(*See figure 24*)

This model (Scale 1 : 48) represents the floating steam fire engine, the *Fire Queen*, which was supplied by Shand, Mason & Co, to the Bristol Corporation for the protection of the docks and waterside property. She was in service until 1922. The pump was a standard *equilibrium* set having three steam cylinders directly coupled to three bucket and plunger type water pumps. The float had a capacity of 1,200 gallons (5,500 l) per minute at 140 lb per sq in (10 kg per sq cm). The vessel was 53 ft (16·2 m) long.

Originally two large hose reels were fitted, but these were replaced in 1900 by the monitor. This is a deck-mounted delivery branch which can be directed mechanically as required and can handle high delivery rates at high pressure, where the reaction on a flexible hose would be too great for a fireman to support.

E

Motor fire appliances

36 Motor Fire Engine 1904
(*See figure 25*)

This machine was the first self-propelled petrol-motor fire engine used by a public fire brigade. It was built for the Finchley Fire Brigade in 1904 by Merryweather & Sons, Ltd, who had built the very first motor fire engine earlier in the year for the Rothschild estate in France. Attempts were made towards the end of the 19th century to improve the transport of fire engines, and several steam propelled machines were constructed with some success. The advantages of the petrol engine, particularly the rapidity with which full power could be obtained after starting the engine, made it specially suitable for this purpose, and before 1904 petrol motors were sometimes used for drawing steam fire engines. This appliance not only used the engine for propulsion but also for pumping and was the first machine to be fully equipped for first aid, as it carried in addition to the main pumps a soda-acid apparatus, hand extinguishers, suction and delivery hose, and a telescopic fire ladder.

The machine had originally a 4-cyl 24/30 hp engine with chain drive to the rear wheels, which gave a road speed of over 20 mph (32 kmph). The pump was driven from the shaft connecting the engine to the gearbox through a sliding pinion, which enabled the pump to be disconnected when desired, and gave a slight speed reduction. The pump is of the *Hatfield* type capable of delivering 250 gallons (1,100 l) per minute. This reciprocating pump has three barrels arranged at angles of $120°$, drawing from a common suction chamber and delivering into a common delivery chamber. Three short connection rods, working off a single crank, operate the plungers. Two hose connections are fitted and a safety valve prevented overloading. The 60-gallon (270 l) soda-acid first-aid cylinder is fitted with three hose connections, a mixing wheel and a safety valve. It was intended to be used until the main pump was in action. The delivery hose and the other gear were carried in a box at the rear of the machine, and the men were carried standing on footboards on each side of the machine.

Since 1904 the machine has undergone several alterations. It originally had a pedal gong, single solid rear tyres, a curved radiator, and a vertical steering column. In 1912 a 50 hp *Aster* 4-cylinder engine was fitted to improve its hill climbing capacity. In 1913 a more modern design of appliance was obtained and the old engine became a stand-by. In 1928 it was sold, the engine and pump were dismantled and used in a gravel pit. It came to the Science Museum in 1930.

In 1966 the appliance was repainted in its original colours, except for the coats of arms which were only cleaned.

37   Motor Fire Engine 1936
(*See figure 26*)

This model (scale 1 : 8) of an open-type appliance represents the *New World* Wagonette fire engine supplied to the Birmingham Fire Brigade by Leyland Motors Ltd. It differs from the earlier fire engines in having the seats facing inwards. The 700 gallon (3,200 l) two-stage centrifugal pump, which is powered by a 115 bhp six cylinder engine, is placed midway between the axles and has suction and delivery branches on both sides. A first-aid apparatus, which suffices to extinguish small fires and is immediately available on arrival, uses a separate gear pump and a reel of 120 ft (37 m) of 1 in (2·54 cm) hose situated behind the front seats.

38   Dual Purpose Appliance 1936
(*See figure 27*)

This is the Dennis open *Big 4* fire engine, which in service carried a wheeled escape. The escape, however, did not come to the Museum. The appliance is powered by a four-cylinder petrol engine developing 90 bhp and there is a two-stage turbine pump. There are two transverse bench-seats for the crew.

This pump-escape was in the service of the London Fire Brigade until 1956.

39   Limousine Type Fire Pump 1936
(*See figure 28*)

For many years the standard type of fire-engine body in this country was the Braidwood, in which the firemen stood or sat facing outwards on longitudinal platforms or seats. On arrival at the fire they could dismount immediately, but were exposed to cold and wet weather and risks of collision on the way. The enclosed body was first constructed for the Edinburgh Fire Brigade in 1930. Enclosed types are now generally used by fire brigades for the protection of their crews.

This model (Scale 1 : 8) represents a Leyland fire engine made for the London Fire Brigade to the designs of the then Chief Officer, Major C. C. B. Morris. The six-cylinder petrol engine developed 114 hp at 2,200 rpm and gave a maximum road speed of about 50 mph (80 Kmph). To increase the pressure from the hydraulic mains, a 2-stage centrifugal pump which could deliver 900 gallons (4,000 l) per minute of water at 85 lb per sq in (6 Kg per sq cm) was fitted. The pump has two suction and four delivery branches at the rear. The priming gear comprises two double-acting piston pumps on the main pump, engaged by a hand-lever. They were used to start a supply from open water and could quickly create a vacuum of 27 in (69 cm) of mercury.

The equipment includes heavily armoured, flexible suction hose with screwed gunmetal couplings, already connected to the pump; several lengths of delivery hose carried in the lockers on each side; a 40 ft (12 m) extension ladder on the roof; several sets, carried inside, of self-contained regenerative breathing apparatus with one hour's supply of oxygen. There is also a first-aid apparatus, enabling the engine to start pumping immediately on arrival at the fire. This comprises a small spur-gear pump driven off the gear-box, a 40 gallon (180 l) tank, and a hose reel at the rear. The full crew consisted of one officer and five firemen.

40 Turntable Appliance 1936
(See figures 29 and 30)

This Leyland-Metz turntable escape, built in 1936, has a four-section ladder which can be extended to 101 ft (30·8 m). The six-cylinder petrol engine drives a 500 gallon (2,270 l) per minute two-stage turbine pump, which can deliver water to the monitor at the top of the ladder, and supplies power for the training, elevation, extension, and plumbing of the ladder, at the base of which are grouped the controls and indicators. The remote control of the engine throttle is brought about by turning a ring which passes round the fulcrum frame. Levers control the motions of the ladder by means of oil-operated clutches. Over-extension, which might endanger the stability of the ladder, is prevented automatically for any degree of elevation. If the ladder is elevated on uneven ground, relative movement between the ladder and a large plumb weight occurs, which opens a valve and actuates the plumbing mechanism to bring the ladder into the vertical plane.

The turntable ladder is employed in a variety of ways, the principal uses being as a water tower and to effect rescues from high buildings, but it can also be used for bridging at very low elevations at right angles to the chassis.

Until 1961 this appliance belonged to the Lancashire Fire Brigade. In 1966 it was repainted and replated.

41 Hydraulic Platform 1965
(See figure 31)

This fire fighting appliance, which is represented by a model to scale 1 : 24, has three articulated booms mounted on a turntable and carrying a cage which will accommodate six adults. It is therefore particularly suitable for rescue work. Water is supplied to the monitor through a $3\frac{1}{2}$ in (8·9 cm) pipe and there is a water-curtain nozzle at the front of the cage to protect the operator, who can control all the motions. There is also a breathing-air supply to the cage. The two main booms are each powered by two hydraulic cylinders and the uppermost boom by a single cylinder. The motions can also be controlled from the ground.

The maximum working height is 85 ft (26 m) and the horizontal reach is 41 ft (12·5 m). The appliance is stable in all positions, but it must first be levelled by the hydraulically operated outrigger jacks.

The appliance is manufactured under the name Simon Snorkel by Simon Engineering Dudley Limited, and can be mounted on various vehicles. This model shows a chassis supplied by E.R.F. Limited. The 65 ft version of the appliance was introduced in 1961.

42   Portable Fire Pump
(*See figure 32*)

This Coventry Climax FWMP pump manufactured in 1966 and exhibited partially sectioned evolved from the first truly portable fire pump produced to Home Office specification in 1951 and derives from the war-time trailer pumps. The letters stand for *feather weight motor pump.*

It weighs 220 lbs (100 Kg) and can be carried by two men over rough ground and into places inaccessible to a wheeled fire engine.

The four-cylinder petrol engine has 750 cc capacity, runs at 3,600 rpm, has magneto ignition, and is started by a rope starter. The power output is 30 bhp and the centrifugal pump delivers 310 gpm (1,400 lpm) at 60 lbs per sq in (4 kg per sq cm). The pump is primed by an exhaust gas ejector nozzle which extracts air from the pump body and suction hose.

43   Fire Boat 1939
(*See figure 33*)

This is a model (Scale 1 : 12) of the *James Braidwood* which was in the service of the London Fire Brigade on the River Thames from 1939 to 1961. The name commemorates the first Superintendent of the London Fire Engine Establishment who was killed at the Tooley Street fire in 1861.

The boat is 45 ft (13.7 m) long and has a draught of 3 ft 6 in (107 cm). She was built by J. Samuel White & Co, of Cowes, and designed for high speed at some sacrifice of output. She has three internal combustion engines each driving a propeller and is capable of a speed of 20 knots. The two outside engines are also used to drive the turbine fire pumps, each of which has a capacity of 750 gallons (3,400 l) per minute at 100 lb per sq in (7 kg per sq cm). The deliveries run to the monitor and to a swivelling delivery head on each side. The intake is through a strainer on each side of the hull. There is also a bypass from each delivery to the intake strainer of the other pump, so that water can be forced under pressure to clear the strainer of any blockage.

Fire Escapes

44 Map of Wivell's Proposed Fire-Escape Stations for London 1836

This map (20 cm x 31 cm) which was engraved in 1836, was prepared by Abraham Wivell, (1786–1849), an artist, when he was trying to increase the provision of fire-escapes in London. It bears the caption: *Mr Wivell's design for placing Fire-escape apparatus at equal distances of half a mile over London and its vicinity.*

Although the formation of a society for the protection of life from fire had been advocated by John Hudson in 1828, the first effective organisation for the purpose was that founded in 1836 as The Fire Association of the South-Western District of St Pancras. In the same year also the Royal Society for the Protection of Life from Fire was established for the purpose of providing fire-escapes throughout the whole of London.

Previously, fire ladders were kept at certain spots by the various parishes, but the inadequate nature of the appliances and the lack of organisation caused the loss of life by fire to be so serious that some improvements were generally demanded. Various portable arrangements for facilitating escape from burning buildings were designed by different inventors; but to Wivell appears to be due the introduction of some of the most important improvements in fire-escapes, and these are embodied in the models (Nos. 45–47) which were used by him about 1836, when lecturing upon this subject. He also advocated the provision of fire-escape stations throughout the London district, as indicated by him on the map, with the result that by 1865 eighty-five of these stations had been established.

45 Fly-Ladder Fire Escape 1836

(*See figure 34*)

This model (Scale 1 : 6) was used by Abraham Wivell to illustrate the lectures on fire escapes of his design which he gave in London in 1836. The model bears the name of the Royal Society for the Protection of Life from Fire and is very complete in its details. It consists of a main ladder, mounted upon a spring-carriage with two large wheels and provided with a lower frame-work by which it can be handled and elevated. The ladder is of such length as ordinarily to reach to the windows of the second floor of a house, while hinged to the top of it is a swinging ladder or fly ladder which, by means of ropes from levers projecting from it, can be swung upwards as an extension that reaches to the third floor. There are also two shorter ladders, one of which is fitted with two hooks so that it can be attached to the folding ladder so as to reach a fourth storey. Beneath the main ladder is a canvas trough, or shoot, down which a person can be passed, and at the upper end of this ladder are rollers to facilitate its elevation against a wall.

This type of fire escape was very generally used in London and elsewhere until about 1880, when the extending type of ladder began to take its place.

46 Extending Ladder Fire Escape 1836
(See figure 34)

This model (Scale 1 : 6) was used by Abraham Wivell to illustrate the lectures on fire escapes of his design, which he gave in London in 1836. The ladder is mounted on a small carriage fitted with two large diameter wheels, so that it can be easily moved to the site of the fire. The ladder is in two sections, the bottom section being fixed to its carriage while the upper portion slides in roller guides, fitted near the top of the fixed ladder, and is raised by ropes passing over pulleys. At the top of the moving ladder a metal frame is fitted having a pulley block for use when lowering persons by a rope.

The extending ladder was not, at first, used so much as the fly-ladder, but it came into general use about 1880.

47 Hinged Ladder Fire Escape 1836
(See figure 34)

This is one of the models (Scale 1 : 6) used by Abraham Wivell to illustrate the lectures on fire escapes which he gave in 1836.

Before 1836 ladders were kept by the various London parishes at certain places for use in case of fire. Wivell's ladder was also intended to be carried to the scene of the fire. It is made in two parts hinged together, so that in its closed form it can be easily transported. The two parts of the ladder can be locked in the extended position by rectangular collars, so forming a long rigid single ladder, when the full length is required.

48 Drawings of Fire Escapes c. 1836

These five drawings, four of which may be by Wivell, show the methods used or proposed by him for saving life from fire.

1 Various forms of life-saving apparatus: most of them employ ropes, but in one case a long rod is used, built up of detachable sections and fitted with wheels at its top end.

At the foot of the drawing is written in ink 'Wivell's public and private fire escapes invented in the years 1827–35'. 44 cm x 53 cm. *(See figure 36)*

2 A fire scene at which a fly-ladder escape (No. 45) and a hinged ladder (No. 47) are in use. 33 cm x 23 cm.

3 A coloured drawing of a fire scene showing two fly-ladder escapes in use. 35 cm x 29 cm. *(See figure 35)*

4 A fireman carrying a ladder on his back supported by straps. 44 cm x 32 cm.

5 A fireman at the top of the main ladder of a fly-ladder escape using an axe. At the foot of the drawing is written 'Drawn by W. Parrott' and 'Mr Wivell in the costume of a fireman, 1836'. 33 cm x 24 cm. *(See figure 37)*

49 Extending Ladder Fire Escape 1908
(See figure 34)

This model (Scale 1 : 6) represents a 50 ft (15 m) Shand Mason lattice-girder extending-ladder fire escape, used by the Guildford Fire Brigade in 1908. For the development of the extending ladder escape, many designs were introduced, in the latter part of the nineteenth century, to increase its strength and safety, to make it more compact, and to reduce its weight. The construction in which the lower ladders were made of roughly rectangular section, so that the upper ladders could be arranged to slide inside them, was patented by James Shand in 1880.

The model shown is a typical design for a ladder for transport on a tender. It is mounted on a frame with wheels for easy movement when at the fire.

The ladder has four telescopic sections. The outer section is fixed to the frame, while the other sections slide one within another. The two outer sections are similarly constructed, having four wooden beams joined at the sides by metal lattice work, in front by the rungs of the ladder, and at the back by bars. The next section is composed of two deep wooden beams joined in front by the ladder rungs and behind by metal bars, while the final, innermost, section is a ladder of normal type.

The ladders are mounted on an unsprung chassis made of angle iron. In order to keep the ladder vertical when the machine is used on a slope, a hand screw, working in a nut on one side of the wheeled axle, enables the angle between the axle and ladders to be adjusted. Stability is further secured by the use of two long props hinged near the top of the bottom ladder and having spikes for sticking into the ground.

At the bottom of the ladder a frame with handles for manoeuvring purposes is fitted. This projects at right angles to the ladder and is automatically locked in position when pulled out.

The ladders are extended by a rope and pulleys worked from a drum near the bottom of the ladder. The drum axle has winding handles on each side of the escape and the ratchet release lever, for lowering the ladder, can also be worked from either side.

Fire Extinguishers

50   Fire Grenades

A fire grenade is a sealed glass flask containing water with chemicals in solution, which was intended to be thrown into a fire with — it was claimed — great extinguishing effect. In some grenades sodium bicarbonate was included, which would evolve carbon dioxide. Grenades were introduced in America in the 1870's and were in vogue for about forty years, but were of little real value in fighting fires. They were kept ready for use in a wire basket or a wire clip hanging from the wall and probably afforded no more protection than the same amount of water in a bucket. The Harden No 1 grenade was patented in 1871 by H. D. Harden of Chicago and the grenade patented in 1883 was sold in Britain by the Harden Star Hand Grenade Extinguisher Company. There are two examples of each in the Collection. The Swift grenade was manufactured by J. H. Heathman & Co. of London.

51   Maintained Pressure Water Extinguisher 1966

The metal container is plastic-lined to prevent corrosion and is charged with two gallons of water. A tube reaches to near the bottom of the container and is connected through the lever-operated valve to a flexible hose and nozzle. The container is pressurized to 150 lbs per sq in (10·5 kg per sq cm) as indicated by the pressure gauge on the head, from a compressed air supply or by using a suitable pump.

This extinguisher, acquired in 1966, is in principle the same as the original extinguisher invented in 1816 by Captain George Manby (1765–1854). (See Cover illustration).

52   Gas Cartridge Water Fire Extinguisher 1956

This extinguisher is operated by striking the plunger, which breaks the sealing disc of a cartridge containing carbon dioxide at high pressure. The CO_2 then expels the water through the flexible hose and nozzle. The container is made of lead-coated steel and is tested to a pressure of 350 lbs (25 kg per sq. cm) A rubber safety valve is fitted near the plunger.

53   Soda Acid Fire Extinguisher 1928

In this extinguisher the gas pressure to expel the liquid is generated by the reaction of sodium bicarbonate solution with sulphuric acid, which produces carbon dioxide. The bicarbonate solution is in the conical metal container and the acid is in a sealed glass bottle which is broken by striking the knob at the top. The bottle is held in a perforated cage and a strainer is fitted to the outlet to prevent broken glass from choking the nozzle.

54 Soda Acid Fire Extinguisher 1954

This extinguisher, shown sectioned, normally contains a solution of sodium bicarbonate in water. When it is inverted the lead weight falls and guided by the rod breaks the bottle containing sulphuric acid. The chemical reaction which then takes place generates carbon dioxide at a pressure which drives water containing neutral sodium sulphate in solution out through the strainer and nozzle. The weight has a ledge which catches on a ball on the rod and is thus prevented from falling until the extinguisher has been completely inverted.

55 Chemical Fire Engine 1902
(See figure 38)

This is a model (Scale 1 : 6) of a horse-drawn chemical fire engine used by the Bristol Fire Brigade. It was originally a manual engine made by Merryweather and Sons in about 1888, but was returned to the makers for conversion. This machine was the first of its kind and was in service from 1902 to 1914. It was used as a first-aid appliance and was in effect a large horse-drawn soda-acid fire extinguisher. The copper cylinder at the rear contained 40 gallons (180 l) of water in which sodium bicarbonate was dissolved. The top fitting on the horizontal cylinder contained a lead bottle, filled with sulphuric acid, which was punctured by turning the handle at the rear. This allowed the acid to mix with the bicarbonate solution. Carbon dioxide gas was generated at a pressure of about 100 lb per sq in (7 kg per sq cm), which expelled the liquid through 120 feet (37 m) of $\frac{3}{4}$ in (19 mm) hose and a $\frac{3}{16}$ in (4·76 mm) nozzle.

The engine operated for about five minutes on one charge and could then be connected to a hydrant to be worked by hydraulic pressure. The third valve on the tank is the flushing valve to give the correct water level. The safety valve was set to blow off at 150 lb per sq in (10·5 kg per sq cm).

56 Chemical Foam Fire Extinguisher 1928

The foam extinguisher is used for smothering fires of flammable liquids such as petrol, in which water will sink, but on which foam will float.
The inner reservoir contains a solution of aluminium sulphate in water and the outer compartment a solution of sodium bicarbonate with a stabiliser such as saponine. When the valve is unscrewed and the extinguisher inverted, a foam consisting of CO_2 gas within the bubbles is expelled through the nozzle by the gas pressure.

57 Maintained Pressure Foam Extinguisher 1966

This type of fire extinguisher is used for fighting fires involving oil, petrol, paint etc. It contains a foam compound dissolved in water and is pressurised by air at 150 lbs per sq in (10·5 kg per sq cm). It is charged by an air pump or from a compressed air line.

The foam-producing action of this extinguisher is mechanical and is brought about by the jet of liquid entering the branch at high speed and entraining air, which is drawn in through the perforations.

The foam floats on the burning liquid and smothers the fire by preventing access of air.

58 Pump-Type CTC Fire Extinguisher 1931

The pump is immersed in the fluid container and is designed to eject the extinguishing fluid, carbon tetrachloride, in whatever position it may be held.

The body of the pump is free to revolve, so that the inlet ports by the action of gravity are always below. The two inlet valves are joined by a sliding rod which closes the upper port and opens the port below the liquid surface. The pump is double-acting and the inlet valves are automatic, the delivery being through the hollow piston rod through ports opened and closed by the piston sliding on the rod. When not in use the pump handle is pushed down and locked in position by giving it a quarter turn. This action also seals the outlet by pressing the delivery pipe against a valve in the handle.

This type of extinguisher was developed in 1925 by the Pyrene Company for use on motor vehicles.

59 CTC Fire Extinguisher 1954

This extinguisher contains carbon tetrachloride pressurised by carbon dioxide at 80 lb. per sq in (5·6 kg per sq cm).

On opening the valve a jet of liquid is expelled which readily vaporises to a heavy vapour which blankets the fire and prevents combustion. Carbon tetrachloride is a non-conductor of electricity and is therefore suitable for use on electrical fires. This type of extinguisher is commonly used on motor cars. As carbon tetrachloride evaporates without trace, it does no damage to fabrics, but exposure to the fumes should be avoided as they are toxic.

60 Automatic CTC Fire Extinguisher

This extinguisher was charged with carbon tetrachloride and pressurised by air at 7 kg per sq cm. It is operated manually by turning a hand wheel. The extinguisher also works automatically when the temperature rises to 65 °C or flames reach a celluloid disc, which causes a fuse to melt and release a valve.

This extinguisher was manufactured until 1940 by Boyce Fabrieken N.V. of Amsterdam.

61 Methyl Bromide Fire Extinguisher 1935

Methyl bromide boils at 4.5 °C to form a heavy vapour which is very effective in inhibiting combustion, which cannot continue if more than 3.5% is present in the atmosphere surrounding the fire. The corresponding figure for carbon tetrachloride is 15%, Methyl bromide is, however, more toxic. Its chief use is for the protection of internal combustion engines.

In this extinguisher the liquid is under pressure in a sealed copper cylinder, which is discharged by striking the plunger.

62 BCF Car Fire Extinguisher 1967

This extinguisher contains 12 ozs (340 gm) of liquid bromochlorodi-fluormethane which discharges under its own vapour pressure to form a smothering blanket of heavy vapour. This substance is practically non-toxic and does not conduct electricity.

The extinguisher is operated by a press button and the discharge lasts for 20 seconds. It must be returned to the manufacturer for recharging.

63 Maintained Pressure Dry Powder Extinguisher 1966

This type of fire extinguisher employs a dry powder extinguishant and is particularly suitable for dealing with fires on the surface of flammable liquids. The chief constituent of the powder is finely divided sodium bicarbonate, of which about three-quarters is finer than 37 microns.

The powder is expelled, on depressing the handle, by a charge of compressed air at 150 lbs per sq in (10 kg per sq cm). It is charged by means of an air pump or from a compressed air line.

64 Pistol-Type Powder Fire Extinguisher 1954

This extinguisher, patented by H. Hutchinson in 1930, is in the form of a pistol and fires a 25 cm long cartridge containing the extinguishing powder. The striker is cocked before firing and released by a trigger. The propellant charge projects a cloud of powder about 6 metres.

65 Dry Powder Fire Extinguisher 1967

This extinguisher, which is intended for home use, is expendable and must be replaced after discharge. The steel canister contains one kilogram of dry powder and is pressurized by nitrogen at 150 lbs per sq in (10 kg per sq cm). It is operated by striking the plastic nozzle cap hard down on to a firm surface to release a jet of powder which flows for about ten seconds with a range of eight feet ($2\frac{1}{2}$ m). The chief constituent of the powder charge is finely divided mono-ammonium phosphate.

66 Carbon Dioxide Fire Extinguisher 1954

This extinguisher contains carbon dioxide stored as a liquid under high pressure. When the valve is opened the CO_2 issues through the funnel as a gas which is heavier than air and tends to blanket the fire on to which it is directed. Air containing more than 17 per cent of CO_2 does not support combustion although oxygen is still present.

The advantage of this type of extinguisher is that CO_2 leaves no trace and does less damage by its impact than a liquid jet. It is therefore suitable for extinguishing small fires in laboratories and for dealing with burning liquids. It is a non-conductor of electricity.

The purpose of the discharge horn is to reduce the velocity of the gas so that it does not act like a blow torch and blow up the fire. It also prevents the entrainment of air with the CO_2.

Fixed Installations

67 Fire Hydrant

This pillar-type hydrant was installed at the Royal Ordnance Factory, Royal Arsenal, Woolwich in about 1890. The hydrant valve is operated by a hand wheel at the top of the column. There are two outlets. The door at the base gave access to a gas jet which was kept alight in the winter months to prevent the water in the hydrant from freezing.

68 Hose Reel

The disadvantage of an ordinary hose is that it must be completely run out and connected with a hydrant before it can be used. The hose reel is mounted on a wall and permanently connected with a rising main. Water enters the rubber hose through a stuffing-box gland and the hollow rotatable shaft. As much tubing as is needed is pulled off the reel through a swivelling ball ring and the nozzle valve is opened when the fire is reached.

69 Carbon Dioxide Fire Extinguishing Installation

This type of equipment is installed for protecting electrical equipment, carding machines, and places where highly flammable liquids are used or stored. The CO_2 is kept as a liquid at high pressure in cylinders which are pierced when a rise in temperature causes a fusible link to melt or by operation of the manual release. The link is included in a phosphor bronze cord anchored at one end and weighted at the other. When the weight falls, it trips the piercing mechanism.

The piercer first makes a hole in the sealing disc and then withdraws to give clear passage to the gas released. The liquid vaporises to produce 450 times its volume of gas, which either distributes generally or is concentrated by a cone nozzle into a particular area. The CO_2 smothers flames by diluting the oxygen in the air and inhibiting combustion, and unlike water does no damage to machinery.

70 Sprinkler Installation
(See figure 40)

Water enters from the town mains through the stop valve and large diameter pipes to lines of 1 in (2·54 cm) pipes running close to the ceilings. Automatic sprinklers are fitted to the small pipes generally at 10 ft ($3\frac{1}{4}$ m) intervals and release a spray of water when a fire starts beneath them.

The system above the alarm valve is charged with water and the pressure keeps the valve closed. Flow of water through a sprinkler head, however, causes the valve to lift and this allows enough water to pass through the annular groove in the valve seating to operate the alarm turbine. Pressure is also applied to the diaphragm in an electric switch and so actuates alarms at a distance. There is a drain to waste from the alarm motor pipe through a restrictive orifice.

A small non-return compensation valve in the alarm valve clack allows water to pass through, if the mains pressure should rise, in order to prevent the main valve from opening and giving a false alarm.

When a fire has been extinguished, the main stop valve is closed to prevent unnecessary water damage and to allow opened sprinklers to be replaced.

Valves to waste are provided for draining the installation and for testing purposes, that on the $\frac{1}{2}$ in (1·27 cm) pipe being equivalent to the opening of one sprinkler.

71   Grinnell Sprinkler Heads $c.$ 1881

The sprinkler head invented in 1881 by Frederick Grinnell (1836–1905), to which minor modifications were made in the subsequent seven years, is exhibited complete, sectioned, and opened by the action of fire. The valve is a soft metal disc held under pressure by a two-lever linkage in a valve seating formed in a flexible diaphragm. The end of the second lever is held to the frame by a soldered joint which melts at 155°F (68°C). When the solder melts, the water is released through the hole and forms a spray on striking the deflector beneath.

As the levers were sometimes found to stick, this sprinkler was superseded in 1890 by an improved type (No. 72).

72   Grinnell Sprinkler Heads 1890

In 1890 Grinnell brought out the improved sprinkler, represented complete, sectioned, and after exposure to fire. The sectioned example dates from 1912 and differs slightly in a few details. In this type of sprinkler a glass valve replaced the metal valve which was liable to stick. It is held in place by a strut composed of three interlocking pieces of metal soldered together. When the solder melts, the pieces, which are under pressure, spring apart and the valve drops out. The water is released, impinges on a deflector and breaks into a spray.

An Atlas sprinkler of very similar construction, dated 1941, is included in the collection.

73   Dowson & Taylor Sprinkler Head 1886

The valve of this sprinkler consists of a thimble-shaped tube soldered inside another tube. A non-conducting material on top of the inner tube keeps the water away from the soldered joint. A distributor drops with the tube and is caught by projections on the body of the sprinkler.

74   Witter Sprinkler Heads 1887, 1898, and 1906

Two sprinklers, one of 1887 and the other of 1898, manufactured by Witter & Son of Bolton, have a valve held in place by a lever hooked at each end to the frame and with an adjusting screw passing through the centre. The frame consists of two levers pivoted at the top and held together at the bottom by solder. The levers fly apart when the solder melts.

The chief improvement in the later sprinkler is the provision of a spiral spring which forces the valve off its seat when it is released.

The E type sprinkler was manufactured in 1914 in accordance with the 1906 patent. The valve is held in place by soldered struts and has the general appearance of the Grinnell sprinkler (No. 72). The valve is of metal and has a spiral spring under it.

75 Hudson Sprinkler Head 1888

This sprinkler was patented by J. C. Hudson and manufactured by Shand, Mason & Co. The valve was held against the seat by a spring-loaded support, which is missing, and a lever, one end of which rests on a fulcrum, the other end being soldered to the frame. When the solder melts, the lever and support fall away and the valve, which has a serrated edge, drops on to the frame and acts as a deflector, causing the water jet to break up into a spray.

76 Titan Sprinkler Heads 1903

These sprinklers intended for upright use were manufactured by George Mills & Co., Ltd. of Manchester. The complete specimen is dated 1920 and the used example is undated.

The valve disc is held in place by a strut consisting of two interlocked pieces joined by a fusible rivet. The other end of the strut is held by a screw in the frame, which carries a small deflector plate.

77 Toggle-Jointed Sprinkler Heads

These sprinklers are similar in having a triangular strut consisting of two levers in contact with each other, one pressing on the valve, the other pressing on the frame, and connected by a fusible link of two flat pieces soldered together.

The International sprinkler was patented in 1903 and this specimen is dated 1909. The Newton B type sprinkler was patented in 1908 and is dated 1920. The Rockwood D type sprinkler was first brought out in 1911. This specimen is marked with a patent date 1930 and was manufactured in 1939. The Atlas D type sprinkler is dated 1949 .

78 Quartzoid Bulb Sprinkler Heads

The usual modern form of sprinkler developed from the Grinnell sprinkler in 1925 is the quartzoid bulb sprinkler which is exhibited both complete and dismantled. The orifice is sealed by a metal gasket held in place by a strong quartzoid bulb containing a highly expansible liquid and a bubble of its vapour. When the temperature rises the liquid expands and the bubble contracts. Eventually the liquid fills the bulb and a slight further expansion increases the pressure sufficiently to shatter the bulb. Seven operating temperatures are available in the range 155°F to 500°F (68–260°C). This type of sprinkler has the advantage that, unlike the earlier metal type, it is not liable to corrosion.

In addition to the bulb sprinklers exhibited, which were manufactured in 1954, the Collection contains a Grinnell sprinkler dated 1927, an Atlas dated 1946, and a Titan dated 1949.

There is also a type of quartzoid bulb sprinkler designed to provide an unobtrusive unit for installation where appearance is an important consideration and where the pipe work can be concealed above a false ceiling. A masking ring conceals the opening in the ceiling.

The sidewall sprinkler is used in corridors and in places where overhead pipework would be troublesome. It is mounted upwards and has an angled deflector plate.

79 Soldered Sprinkler Head, 1966

A modern form of soldered sprinkler is used instead of the quartzoid bulb sprinkler in premises such as food-processing plants where the scattering of quartzoid particles when a sprinkler opens would be undesirable. The solder is almost completely enclosed and has a protective film against corrosion.

80 Water-Spray Projector Systems

The *Mulsifyre* System is designed to protect premises containing those flammable liquids, such as oil, which are not miscible with water. Water passing through the projector at a pressure of at least 40 lbs per sq in (2·8 kg per sq cm) issues in the form of an expanding cone of finely divided broken streams of water travelling at high velocity, and on striking the burning liquid forms with it an emulsion which is non-flammable. Two projectors, which are nozzles of special design, and sectioned examples are included in the exhibit.

Groups of projectors are connected to the water supply through a control valve, which opens when a predetermined temperature is reached. The control valve is maintained in the closed position by means of a frangible quartzoid bulb, the action of which is described in No. 78. A complete and a partially sectioned control valve are exhibited.

The protectosprayer, two examples of which are shown, discharges a cone of medium size water droplets and is used for cooling tanks containing flammable liquids and gases. It can be operated both manually and automatically.

81 Drenchers

A drencher installation provides a discharge of water over the external openings of a building. It comprises a system of pipework fitted on the outside of a building with drencher nozzles at suitable intervals. Drenchers are also used in theatres to protect the safety curtain.

The spray may be released automatically by a quartzoid bulb or manually by a drencher stop valve.

A wall or curtain drencher and a window drencher are exhibited.

F

Fire Alarms

82 Fire Detector 1878

This device invented about 1878 by J. Preece is intended to complete an electric circuit, and thus give the alarm, when the temperature of the surrounding air exceeds a certain value.

The mechanism consists of a short brass cylinder with a terminal at its closed end. Into the other end is screwed an insulating plug through which passes a brass rod forming the other terminal and having fixed to it the moving contact, which is situated within the body of the cylinder. The fixed contact is screwed into the side of the cylinder and is adjustable. The moving contact is a flat spiral spring composed of two strips of metal having different coefficients of expansion, soldered together so that the metal having the greater coefficient is on the inside. As the temperature rises the coil unrolls until its free end touches the fixed contact point thus completing the circuit. A portion of the cylinder is cut away to facilitate the adjustment of the contacts.

83 Fire Detector 1884

This is an arrangement, patented by G. L. Pearson in 1884, by which an electric bell is rung when the temperature in any selected positions in a building rises above a certain fixed degree. It consists of a number of thermometers distributed over the rooms that are to be protected, connected in parallel by two insulated wires that continue the circuit through a dry battery, a call bell, and an indicator board. Each thermometer has fused into its tube, at a height corresponding with the danger temperature, two platinum wires so that when the mercury rises sufficiently it places them in electrical contact, and thus closes the bell circuit.

84 Street Fire Alarm 1910

This call box was one of some two hundred such alarms connected by telegraph wires with apparatus in the chief fire station in Manchester. It was installed by the Gamewell Company in 1910 and was in service until 1951. Such alarm systems are not now used in this country, as it is considered that enough telephones are available for notifying outbreaks of fire.

A battery at the station sends a continuous current through the alarm circuit. When the handle is turned a spring is first put under tension and then the mechanism is released at a certain point and makes a series of disconnections which signals the number of the box to the station. In the switchroom the signals sound a gong, light a lamp, and indicate the number of the call box on the annunciator panel, and record the number and time of call on the paper tape of the pen register or on the punched tape register. A repeater relays the signal to gongs and indicators in the sub stations.

A bell in the street alarm also rings in order to draw attention to its use, in the hope of deterring malicious callers.

85   Pearson A Type Fire Detector 1907

This detector was patented in 1906, but was superseded by the B.1 detector in the following year.

The fire alarm circuit is completed by the movement of a contact carried by a thin bimetallic strip against an adjustable limit stop, which constitutes the other contact. The curvature of the strip increases as the temperature rises until a certain pre-set temperature has been attained, and then the detector completes the alarm circuit.

86   Original Pearson B1 Fire Detector 1907

This is the original experimental B.1 fire detector used by its designer, Alfred Henry McNeil, which superseded the A type detector. Two curved bimetallic strips of unequal thickness, which are attached at the ends to a rigid base but insulated from each other, expand when heated with the result that their curvature increases. If heating is rapid the thinner strip expands more rapidly than the thicker one and makes contact with the screw attached to its centre, thus closing an alarm circuit connected across the strips. If the heating is slow the strips expand equally so that the gap is maintained. In this way the detector is rendered sensitive to an outbreak of fire but does not respond to normal fluctuations in temperature. The sensitivity depends on the setting of the screw.

87   Pearson B1 Fire Detector and Alarm 1955

This fire detector is the same in principle as the original B.1 detector, (No. 86). The two strips are protected by a grille and carry the contacts on the ends of rods which pass into the body of the apparatus, the top cover of which has been replaced by perspex for the purpose of exhibition. The contact attached to the thicker strip can be raised or lowered by means of a screw to be seen at the centre of the lower circular scale. This adjusts the sensitivity. It is normally set so that the contacts close for a quick rise in temperature of 40°F (22°C) above room temperature. A refinement in this detector is provided by the adjustable limit stop, which halts the thicker strip when a certain temperature is attained, however slowly. The normal setting of the limit stop is 180°F (82°C).

Detectors are installed in suitable positions in the building and connected with the indicator panel. Should a fire occur where the detector is situated, the circuit is closed. Each detector is connected through its corresponding drop-flap annunciator to the three relays in series at the top of the panel. One relay operates the warning bell which continues to ring while any detector is in action. The annunciator shows where the fire is. The other two relays reverse the connections of a dry battery, of which one pole is earthed and the other connected with the fire station through a telephone line.

In the fire station the line from each building enters the corresponding indicator panel and is connected with a relay in series with a dry battery, which is earthed. In the normal condition the batteries in the building and in the station are in opposition so that no current flows. The reversal of the building battery operates the relay, which rings a bell and drops the annunciator flap, showing which building is giving the alarm. The bell continues to ring until the operator on duty pushes the plug into the verifying jack. This action disconnects the relay and substitutes a voltmeter which reads 'FIRE' when the combined battery voltage is across it. The purpose of this procedure is to discriminate against false alarms. Earthing of the telephone line, for example, would cause the station battery alone to ring the bell and drop the flap, but would give a 'LINE FAULT' reading on the verifier meter. The operator would then telephone to ascertain the cause. If the test key is pressed the station battery is eliminated so that the building battery operates the alarm, if the system is in order.

88   National Tubular Fire Detector 1955

This type of detector utilises the expansion, when heated, of a liquid contained in a thin copper tube, which is fixed along the ceiling mouldings. It was designed for use in situations where a visible installation would be undesirable.

The tube is connected to two metal bellows, which support contact arms. A slow rise in temperature caused by normal heating moves both diaphragms equally. Quick heating moves one diaphragm more rapidly than the other, to which flow is restricted by a very small inlet, causing the one contact-arm to overtake the other, thus closing the alarm circuit. Slow heating to a dangerous degree causes the contacts to meet the fixed limit contact, also closing the alarm circuit.

To test the apparatus the lever of the three-way cock is turned $90°$ anti-clockwise so as to put the rapid diaphragm into communication with the pump and to seal off the compensating slow diaphragm. The pump key is then screwed until contact is made between the contact arms.

89   Fixed Temperature Fire Detector 1966

In some situations such as a boiler house, where a rapid rise of temperature may be a normal occurrence, a rate of rise detector (Nos. 86-88, 90). is unsuitable, and a fixed temperature detector is preferred. This example consists of a hollow cylindrical metal electrode with fins to collect heat Inside is a central rod electrode passing through a supporting polythene plug and insulated from the cylinder by the air gap. The cylinder is lined with a low temperature melting alloy. When this melts, the electrodes are connected and the alarm circuit completed.

With the detector is exhibited a twice-size sectioned model.

90   Rate of Rise Fire Detector 1966

The electrodes are attached to two coiled strips cut from a bimetallic sheet, mounted in a metal and plastic housing screwed to the ceiling. The strips bend inwards when the temperature rises. If the rise occurs rapidly, the upper strip being shielded by the thick plastic cover, is over-taken by the lower strip, which heats up more quickly; contact is made and the alarm is given. The strips respond equally, however, to slow daily and seasonal temperature changes and no contact is made, unless the upper strip come up to the stop.

The alarm is thus given when the ambient temperature rises rapidly, or even if slowly, to a predetermined temperature.

The detector is included in the alarm installation exhibit (No. 91) and there is also a twice-size sectioned model.

91   Fire Alarm System 1966

The Pyrene system employs fire detectors of the rate of rise of temperature type (No. 90) or, in some instances, of the fixed temperature type (No. 89). The detectors are connected with a control unit which switches on fire bells, indicates visually where the outbreak has occurred, and transmits a signal by telephone line to the local fire station or to a fire report room. In London fire alarm installations are connected with the watch room maintained by Associated Fire Alarms Limited, whence fire calls are passed on to Fire Brigade Headquarters.

The control unit is operated by the AC mains, but a stand-by battery comes into service automatically in the event of a mains failure.

The circuits are arranged to give warning automatically should any part of the system fail.

92   Ionisation Fire Detector and Alarm Installation 1966

This detector is sensitive to the presence in the air of ionised particles resulting from combustion even before flame has broken out. It was invented in 1941 in Switzerland by W. Jaeger and E. Meili.

In the sectioned detector are to be seen two ionisation chambers one above the other and part of the glass envelope of the cold cathode relay tube. Both chambers are rendered conducting by exposure to a radioactive source and are connected to form a voltage divider across a 216 volt supply, with the common point connected to the trigger of the tube, which is also across the supply.

The lower chamber is open to the air, so the presence of combustion products affects its conductivity and causes the potential of the common point to rise by about 27 volts. A fairly heavy current then flows through the tube and actuates relays in the control unit.

This detector forms part of the Minerva Fire Alarm System. The control unit is powered by a battery, trickle-charged from the mains, and, as in No. 91 sounds alarm bells, give a visual indication of the location of the fire, and signals the fire station or the fire report centre.

Miscellaneous

93 Leather Fire Buckets
(*See figure 1*)

Four leather buckets: one with the mark of the Sun Fire Office and two with the initials FP and unicorn crest.

94 Fire Engine Book
(*See figure 41*)

The first book on fire engines was written by Jan van der Heiden and his son Jan, and published in 1690. This is the second printing which appeared in 1735. The short title in Dutch is Slang-Brand-Spuiten.

Jan van der Heiden (1637–1712) was a distinguished artist and also Superintendent of the Amsterdam Fire Brigade. He introduced the use of the delivery hose, made of leather, an invention which made the fire engine much more effective than it had been with the swivelling goose-neck delivery jet, by enabling fires to be attacked at close quarters. The cistern was connected by flexible hose with a canvas trough in a wooden stand which was sited on a canal bank and kept filled with water raised in buckets. This obviated the use of a long bucket chain. Van der Heiden later introduced the wired suction hose, which dipped into the canal. He was the illustrator as well as the author of this book, which is exhibited open at the plate showing the fire in 1673 in Amsterdam at which the new pumps were used for the first time and the old pumps for the last time.

There are 25 plates, which are captioned in Dutch and in French.

95 Fire Hose

Leather hose at first had sewn seams; but riveted hose, which is less liable to leak, was patented in 1818 by Pennock and Sellers of Philadelphia.

This specimen of riveted leather hose with screw couplings at each end is part of the equipment of the Welbeck engine (No. 17). There is also a hose bandage or clamp of leather with screw fastenings for making temporary repairs.

Examples of plain canvas hose, rubber-lined canvas hose, and canvas suction hose wired to resist external pressure, date from 1924. An exhibit on the development of hose includes specimens of hose of:– leather; flax (1900); rubber-lined cotton (1930); reinforced rubber-lined cotton/nylon (1943); cotton warp, synthetic weft, reinforced rubber-lined (1947); reinforced rubber-lined PVC coated all-synthetic jacket (1958); and all-synthetic jacket, lined and coated with synthetic rubber (1965).

96 Couplings & Nozzles

A standpipe and branchpipe were part of the equipment of the Welbeck brigade (No. 17).

Examples of screw couplings and instantaneous self-locking coupling for connecting lengths of hose were acquired in 1924. The latter are held together by spring latches.

A coupling with a design of leather-seal which tightens under pressure, patented by Tribe and Hele Shaw in 1920 and 1921, dates from 1928.

A combination nozzle, dating from 1924, has a rotatable valve which can be used to cut off the water supply or connect it with a small first-aid jet or with the main jet. It can also be adjusted to give a variable diffused jet.

97 Hose Reel 1866

This appliance is quite distinct from the apparatus fixed to a wall (No. 68), which the name now signifies.

It consists of a wooden reel mounted between two carriage wheels and hand-drawn by a drag handle. It was used for transporting hose. A box mounted above the reel carried a standpipe, branchpipes, and tools.

This example was supplied by Merryweather & Sons to the Duke of Portland's Welbeck Abbey Estate.

98 Fire Marks and Plates

(*See figure 39*)

The Great Fire of London in 1666 led to the founding of fire insurance companies, which soon formed their own fire brigades.

Up to the first quarter of the nineteenth century the greater part of organized fire fighting was carried out by these companies. To distinguish the houses which they had insured against loss by fire, each company fixed its own leaden fire mark on the front of its houses. The company's firemen then recognized the fire mark, proceeded to fight the fire and sometimes refused their aid if the house did not carry the company's mark. The companies and dates of the marks exhibited are:

    The Sun Fire Office. 1720. Policy 23055. (Established 1710).

    The Sun Fire Office. 1733. Policy 60961.

    The Hand in Hand Fire and Life Insurance Society. 1773. Policy 88635 (Established 1696).

    The Royal Insurance Company. (Established 1845).

    The County Fire Office. *c.* 1860. (Established 1807).

The first of these has been restored to its original colouring: with gilded sun face and number and sky-biue background.

The numbers on the fire-marks are the numbers of the policies. Fire plates, which are unnumbered, largely replaced marks early in the nineteenth century. Their chief purpose was for advertising the company.

99 Firemen's Helmets

The early firemen's helmets were of leather. That included in the group of helmets is from Welbeck (No. 17). The comb is decorated with a lion's head medallion.

Chief Officer Eyre Massey Shaw introduced the brass helmet into the Metropolitan Fire Brigade in 1866 and they were soon adopted by other brigades. The example shown is from Hastings. Brass helmets were later found to be dangerous owing to the possibility of coming into contact with bared electric wiring at fires and in 1936 the London Fire Brigade adopted the cork helmet.

The modern helmet (1966) is constructed of three-ply cork, covered with flame-proofed cotton and finished with acid-resisting and heat-resisting enamel.

100 Fireman's Smoke Helmet late 19th Century

This helmet is made in accordance with S. Barton's patent of 1872.

The helmet is made of light material impregnated with rubber to render it gas-tight and is fitted with glass to protect the eyes from the effects of fumes. It is supplied with a quantitiy of loose material about the neck which may be tucked into the clothing to make an air-tight joint. Within the helmet is a mouthpiece fitting closely round the mouth and, in breathing, the air is drawn in through a filter fitted directly in front and is discharged through a valve on the left side of the helmet. For use in smoke the inventor recommended the use in the filter of alternate layers of charcoal and cotton wool or other fibrous material saturated with glycerine and, if carbon dioxide was expected to be present in considerable quantity, the addition of granulated quicklime was advised.

101 Portrait of P. R. Hodge

Oil painting, 52 cm x 39 cm, by Goldsworthy of Paul Rapsey Hodge 1808–c. 1870. Hodge, who was born in St Austell, built the first self-propelled steam fire engine (No. 23) at his works in New York. From 1850 he was active as an inventor in Britain.

A Select Bibliography

Braidwood, James. *On the Construction of Fire-Engines and Apparatus, the Training of Firemen, and the Method of Proceeding in Cases of Fire.* Edinburgh. 1830.

Braidwood, James. *Fire Prevention and Fire Extinction.* London, 1886.

Young, Charles F. T. *Fires, Fire Engines, and Fire Brigades.* London. 1866

Shaw, Eyre M. *Fire Protection. A complete manual of the organisation, machinery, discipline, and general working of the Fire Brigade of London.* London. 1890.

Sachs, Edwin O. *A record of the International Fire Exhibition, Earls Court, London, 1903.* London. 1904.

Gamble, Sidney Gompertz. *Outbreaks of Fire. Their causes and means of prevention.* London. 1931.

Dana, Gorham. *Automatic Sprinkler Protection.* New York. 1919.

Home Office (Fire Service Department). *Manual of Firemanship.* Seven parts in nine volumes. London. 1943–65.

Blackstone, G. V. *A History of the British Fire Service.* London. 1957.

Jackson, W. Eric. *London's Fire Brigades.* London, 1966.

Fire Protection Year Book and Directory. London. Annually.

Inventory and Photograph Numbers Donors and Lenders

Catalogue Numbers	Inventory Numbers	Photograph Numbers	Donors and Lenders
1	1877–421	752/53 535/67	Vestry of St Dionis – Backchurch
2	1959–312	535/67	Provost and Fellows of Eton College
3	1960–81	535/67	J. F. Bruton
4	1861–59	751/53	H. H. Howell
5	1948–337	545/49	George H. Gabb Bequest
6	1875–5	682/66 683/66	Town Council of Dartmouth
7	1953–371	551/49	
8	1920–122	469/37	Lt-Col B. G. Way
9	1948–400	112/66 (back) 113/66	George H. Gabb Bequest
10	1962–142		The Newcomen Society
11	1964–282	536/67	
12	1953–428	16/54	
13	1914–399	1227/54	H.M. Office of Works
14	1932–269	1097/54	
15	1938–484	36/47	
16	1943–85	657/53	Miss S. Spindelow
17	1958–135, –136	676/59	Nottinghamshire Fire Brigade
18	1894–154	30,885 (detail) 54/66	
19	1918–195	584/37–587/37	Col J. D. K. Restler
20	1965–76	537/67	Royal Ordnance Factory, Woolwich
21	1956–90	674/59	Ministry of Works
22	1869–23	25,088	
23	1869–24	25,089	
24	1924–211	621/66 622/66	Merryweather & Sons Ltd
25	1936–316	747/53–749/53	W. Spindelow
26	1918–194	580/37–583/37	Col J. D. K. Restler
27	1891–168	30,442	
28	1891–169	556/49	

Catalogue Numbers	Inventory Numbers	Photograph Numbers	Donors and Lenders
29	1892–128	30,065	
30	1892–129	555/49	
31	1959–109		G. H. Hindson Bequest
32	1945–3	557/66	Borough of Southgate
33	1920–21	5989, 5991, 541/67, 542/67	London Fire Brigade
34	1966–287		H. B. Thomas
35	1959–108	623/66	G. H. Hindson Bequest
36	1930–256	555/66, 556/66	
37	1939–20	750/53	Leyland Motors Ltd
38	1956–204		London Fire Brigade
39	1936–384	48/54	Leyland Motors Ltd
40	1962–57	86/62, 405/66	
41	1968–129	28/68, 29/68	
42	1966–96	558/66	Coventry Climax Engines Ltd
43	1965–58	691/66	J. Samuel White & Co Ltd
44	1899–74		
45	1899–74 1899–128, –129	23/54	
46	1899–127		
47	1899–74	24/54	
48 (1)	1903–134	20/54	Woodcroft Bequest
(2)	1903–135		Woodcroft Bequest
(3)	1903–136	22/54	Woodcroft Bequest
(4)	1903–133	21/54	Woodcroft Bequest
(5)	1903–132	19/54	Woodcroft Bequest
49	1934–405	25/54	G. H. Hindson Bequest
50	1967–145, –146 1955–118		D.K.A. Durling G. B. L. Wilson
51	1966–326		The Pyrene Co Ltd
52	1954–646		The Pyrene Co Ltd
53	1928–1080		J. Blakeborough & Sons Ltd
54	1954–443		Antifyre Limited
55	1946–92	325/51	G. H. Hindson Bequest
56	1928–1079		J. Blakeborough & Sons Ltd
57	1966–328		The Pyrene Co Ltd
58	1931–110		The Pyrene Co Ltd
59	1954–444		Antifyre Ltd
60	1967–48	831/66	The Fire Protection Association
61	1935–167		The National Fire Protection Co Ltd
62	1967–32		The Pyrene Co Ltd

Catalogue Numbers	Inventory Numbers	Photograph Numbers	Donors and Lenders
63	1966–327		The Pyrene Co Ltd
64	1954–440		Antifyre Ltd
65	1967–31		The Pyrene Co Ltd
66	1954–442		Antifyre Ltd
67	1965–74		Royal Ordanance Factory, Woolwich
68	1954–445		Antifyre Ltd
69	1966–158		The Pyrene Co Ltd
70	1962–364	438/67	Mather & Platt Ltd
71	1950–126 1890–115 1950–127 1967–442		P. G. Woodhouse Dowson, Taylor & Co P. G. Woodhouse M. Butler
72	1966–356 1912–117 1890–115 1967–449		Mather & Platt Ltd Mather & Platt Ltd Dowson, Taylor & Co M. Butler
73	1950–125		P. G. Woodhouse
74	1950–123, –124 1967–448		P. G. Woodhouse M. Butler
75	1967–453		M. Butler
76	1967–443, –444		M. Butler
77	1967–446, –452, –447, –451		M. Butler M. Butler
78	1954–110, –111 1967–441, –450, –445 1966–359, –315		Mather & Platt Ltd M. Butler Mather & Platt Ltd
79	1966–314		Mather & Platt Ltd
80	1954–112, –113 1966–317, –357, –358, –391		Mather & Platt Ltd Mather & Platt Ltd Mather & Platt Ltd
81	1966–318		Mather & Platt Ltd
82	1923–309		Killingworth Hedges
83	1896–140		Pearson's Automatic Fire Indicator Co
84	1952–201		Manchester Fire Brigade
85	1954–657		Associated Fire Alarms Ltd
86	1954–144		Associated Fire Alarms Ltd
87	1954–658 1955–30		Associated Fire Alarms Ltd Associated Fire Alarms Ltd
88	1955–29		Associated Fire Alarms Ltd
89	1967–59 1966–3		The Pyrene Co Ltd The Pyrene Co Ltd
90	1966–2		The Pyrene Co Ltd
91	1965–134		The Pyrene Co Ltd
92	1965–135		The Minerva Detector Co Ltd

Catalogue Numbers	Inventory Numbers	Photograph Numbers	Donors and Lenders
93	1920–141	535/66	
	1959–110		G. H. Hindson Bequest
94	1952–41	655/53, 7888	
		615/68 - 617/68	
95	1924–215, –216,		Merryweather & Sons Ltd
	–217		Merryweather & Sons Ltd
	1958–137, –146,		Nottinghamshire Fire Brigade
	1967–147 to —155,		George Angus & Co Ltd
	1968–430		George Angus & Co Ltd
96	1958–144, –147		Nottinghamshire Fire Brigade
	1924–212, –213, –214		Merryweather & Sons Ltd
	1928–1014		The Victaulic Co Ltd
97	1958–145	675/59	Nottinghamshire Fire Brigade
98	1939–192	539/67	The Sun Insurance Office Ltd
	1938–181		
99	1958–136		Nottinghamshire Fire Brigade
	1966–216		
	1966–151		London Fire Brigade
100	1930–907		The Duke of Devonshire
101	1884–213		

Index

Printed in England for Her Majesty's Stationery Office
by Eyre and Spottiswoode Ltd Grosvenor Press Portsmouth
Dd.138226   K24